THIS AIN'T THE BEER THAT YOU'RE USED TO

A BEGINNER'S GUIDE TO GOOD BEER

DOM "DOOCHIE" COOK

BROKEN WORLD PUBLISHING

THIS AIN'T THE BEER THAT YOU'RE USED TO
Dom "Doochie" Cook

A Broken World Publication
PO Box 11643
Portland, OR 97211
This Ain't The Beer That You're Used To
Copyright © 2019 by Dom "Doochie" Cook
ISBN 978-1-949410-13-6 (ebook);
ISBN 978-1-949410-14-3 (paperback)

Cover Design: Latiesha Cook
Cover Photo & Other Photographs: Russell Breslow
Edited & Formatted by: Christopher Barnes (Cissell Ink), Certified Cicerone® &
MBAA Beer Steward

CONTENTS

DEDICATION

I'd like to dedicate this book to my wife, Tiesh; you've held me down for close to two decades. You've always reminded me that I had more to offer, and when times get hard, you always buckle up for the ride. You've been ten toes down since the beginning, and for that, I'm thankful. I also would like to dedicate this book to my children who have constantly told me that I'm a "beer king" and for being patient and understanding as I've had to dedicate so much time to work and study. My prayer is that my example will inspire you all to strive for anything you want to achieve. All things are possible. Baby Josh, this book is for you, and your name will always be remembered.

I also would like to dedicate this book to my brother, D. Sparks. My last memory with you was drinking a Guinness in the building lobby. It pains me that we didn't get to spend more time together, but I pray that you're resting in peace, moe.

To the beer industry, thank you for the fire that you constantly light beneath me that pushes me to push back harder for change. If I was comfortable, this book or any of the work that we do wouldn't have been a thought. Oz! Good looking for your help! You're a real one, and I appreciate you!

And lastly, this book is dedicated to everyone who looks like me. Yes, I want you to know about and enjoy great beer. I also want you to know that no matter where you start it doesn't have to dictate where you finish and that you can get the job done through hard work and integrity all while being true to yourself. And in the process, you can use whatever you have available to spark positive change in the world around you.

ACKNOWLEDGMENTS

Credit is due and given to the Cicerone® organization, as well as the Beer Judge Certification Program, you both have given me the keys to education that has helped me and will continue to help me help others. Last but not least, shout out to Stephan Mansfield for taking the time to write the story of Guinness that changed my life.

1

BEER IS NASTY AND CHEAP

Picture this, a cold winter night in the Soundview section of the Bronx. A group of adolescents clothed in North Face and Sean John snorkels over top of Champion hoodies. Baggy jeans and Timbs piled up in a local project building lobby with no purpose in mind, just living. This was every day for us, winter, summer, spring, fall. It was a way of life, and one thing that was a constant in that daily congregation was, you guessed it, beer. It was usually ten or so 40-ounce bottles of 211, a nasty-ass cheap malt liquor that had a fierce reputation to leave you faded fast. Here and there you'd find an Olde English, Colt 45, St. Ides, but usually, it was the trusted and proven good, old 211.

We drank every day, no, not because we loved the taste. Nah, it was because we had no direction and only wanted to hang out and get faded, and 211 did the job and quick! We weren't old enough to buy beer and a lot of times we spent our money on other things like weed and getting fresh so when it came time to get beer we boosted it more times than not. No, not stealing in the sense of sneaking it, one of us would brazenly walk into the bodega and walk out with an arm full of 40s, even making multiple trips, daring the employees to say something. Not cool, but this is what it was.

As a youngin', this is how we moved and how we acquired most of the beer we indulged in, but everyone else we knew who indulged actually paid for theirs. I mean, a 40oz beer that's 8% alcohol by volume (ABV) for $2.00? Anyone could buy it – even crackheads could hustle enough change to buy a couple a day. Taste what? Taste who? When your only desire is to get twisted who cares about the flavor? Back then, I know I didn't; but now, I damn sure do. These days, I also care about paying for my beer.

Cheap, high gravity beer has been in the inner cities for decades. Conveniently targeting African-Americans and other minorities through physical advertising as well as paid TV and radio ads and programming using the likes of Redd Foxx, Billy Dee Williams, and many other "models" of color. It's important to note, malt liquor didn't start off this way. It was first marketed to middle-class whites as a premium beer at a higher cost, but they didn't fall for it. Only after realizing they couldn't sell it to their intended audience and that blacks were already buying it did they set their sights on us. If you're old enough, you may remember these ads. But there was a new player on the horizon, plotting out their takeover of this area of the beer industry and the plan was genius. A new, cool, hip, yet super focused set of advertisements from St. Ides malt liquor in the late eighties, early nineties using the likes of Wu-Tang Clan, MC Eight, Ice Cube and tons of other highly influential hip-hop artist of the time took malt liquor to a level where it hadn't been before.

Using hip hop to sell a shit load of nasty beer was an idea birthed by McKenzie River Corporation, a beverage marketing firm owned by marketing wiz Minott Wessinger whose grandfather had his hands in Olde English (OE). Wessinger's company also coincidentally released 211 malt liquor, go figure. They used a pocket full of money and genius advertising to attract the urban demographic and lock their competition out by selling cheap malt liquor and flavored non-alcoholic drinks like Crooked I, promoted by the hottest hip-hop artists of the time. Not that they had to worry about too much competition in the first place, since the movement of good, high quality, flavorful beer didn't have the urban demographic on their mind.

This was during one of the first booms of the craft beer industry which got its start (even if a false start) in the '70s. The founders of this movement had grown tired of the yellow, tasteless beer that was being sold to them, and through the likes of people like Michael Jackson (a legendary beer writer-not the legendary pop singer), they were exposed to an array of new beer styles. 1978 marked a new beginning with the changing of a law in America that allowed people to start brewing beer at home (home brewing). These two things set the tone for new visionaries with dreams of brewing beer professionally to open up breweries serving up the freshest, most flavorful beer they could brew. The only catch was, the urban market wasn't their focus.

But the urban market was somebody's focus, it was an open

playing field, and who better to cash in on that goldmine than a marketing firm who had an advanced understanding of how hip-hop and marketing worked coupled with the money and connections to get shit done? They knew we only wanted a quick buzz. They hit the fucking jackpot by using our new influencers to sell us a fast buzz cheap. Looking at the commercials, you would have thought St. Ides was owned by a black man – definitely not the case. Wessinger used Hip-Hop to sell bullshit to the inner cities while handing out bags to these newly famous rappers to jump on board, but not all rappers were with the shits. Rapper Chuck D of Public Enemy sued McKenzie River Corporation for illegally using him in an ad. I'm sure you'll be shocked at what happened. The corporation stepped away from the situation stating that the campaign was run independently. Guess who was in charge of making and running the ads? DJ Pooh.

McKenzie River Corporation eventually cashed out for hundreds of millions by selling different brands under their umbrella to different corporations. They came in, got that money while further helping to ruin many people's view of beer and then dipped out. But the influence and lasting effects of cheap, nasty beer, both high and low gravity continues to taint urban communities across the country.

It was normal to see 40-ounce beers in your favorite hood flicks, favorite videos, and to hear your favorite rapper talking about drinking a forty. It was rare to hear anyone mention a solid beer brand unless you'd read "The Spook Who Sat By The Door" and saw how Freeman enjoyed his Ballantine Ale and even then you'd probably thought to yourself, "What the fuck is that?" What was rampant, besides malt liquor, was the promotion of high-end liquor and champagne which helped cement into our minds that beer was lesser than—no matter how unknowingly wrong our favorite artists were.

This message carried on an age-old lie that beer was for the poor, everyday working man only, and champagne and wine were reserved for the wealthy, a narrative that has been around forever. What we don't hear is the history of beer and all that it has contributed to our world and lives whether directly or indirectly. Don't worry though, we'll touch on some of this in a bit.

As we got older, our beer habits and tastes changed although not for the better. We went from drinking cheap malt liquor to drinking cheap imported lagers. Coronas and Heinekens were the wave of this age, served in clear or green bottles with aromas that, if you really paid attention, would remind you of a skunk. We loved killing skunked beer, and we ain't even know it. These beers, although slightly more expensive, still painted a lowly picture of what beer was.

Mass marketed and mass-produced beers that are perfect in execution, but poor examples of what beer really is in the full scope of things, that was the vibe that we were on. Before we continue, let me just make it clear that mass market doesn't automatically mean something is trash, just like limited doesn't mean something is of a higher quality, but the mass market beers that we drank were definitely trash-in-taste, specifically. We drank them on the block, at games, birthday parties, at cookouts, while playing spades and running Boston's on our opponents, while playing dominoes, or rolling ceelo; there wasn't a time that we didn't have a Heineken or Corona in arms reach.

Nothing has changed, even now you will see ice cold Heinekens, Stellas, and Coronas with limes being downed while everyone hangs out having a great time. And even in 2019, you can see that Colt 45 now carries the same torch that St. Ides did in the nineties. Just take a stroll through their social media, you'll see some of the biggest and most well-known urban and hip-hop artists and influencers plastered throughout. Add on the "thot juice" like Bud Light's lime-a-Rita and all the other fruit forward malt beverages, and it's the same story repeating itself as history does if we choose not to learn from it.

Look, if you love these beers, I'm not here to knock you, do you. I'm simply aiming to show you that more is available. Malt liquor and imported lagers are two different types of beer, but one who is not well versed in beer would never know. One bad night with a 211 will turn you off concerning all beer, because really, what's the difference in beer? Once you've tasted one Corona you've tasted all the beers, right? All beer is nasty and cheap and tastes the same. You know, they all have that "beer taste," and this is where I have to draw the line because these beers do not represent beer as a whole, no, not by a long shot. Trust me; as we explore and talk beer, you will realize that this ain't the beer that you're used to.

2

HOW I CAME TO LOVE BEER

My introduction to good beer wasn't intentional, meaning, I didn't seek it out. Neither did anyone tell me the "good news" of good beer. I didn't even know that there was such a thing as beer that was superior in taste compared to what we were brought up on. My introduction was actually the opposite, I stumbled upon it by chance. I guess it was fate. I came up in a rough neighborhood, in a rough city during a rough era, surrounded by cheap, flavorless beer; it was obvious good beer wasn't intended for me – until it was.

When I stumbled upon good beer, I was seven years sober. A teetotaler is what they called it in the old days. I was actually opposed to drinking – even in moderation. A lot of us grew up in similar ways, and when everyone around you is growing up the same way, it all seems normal. Now as a father raising kids who are totally different than I was, I recognize how extreme my younger years were. It was that extremeness that eventually lead me to another extreme—religion.

Like I said, I grew up in a pretty rough era, and like any other young, misdirected kid in the hood, I had my hands in a lot of bullshit. School wasn't a factor for us, not even in the slightest. Who was going to go to school and make it big? We would rather play the block for most of the day and spend what little time we had left in the studio chasing rap dreams.

Fresh out of the 8th grade I was accepted into a well-known all boys catholic school named Rice that sat in the middle of Harlem. Sadly, this school is no longer around, but back then, our name rang bells for our basketball team who was well known all over for the caliber of players we put out and the lit ass parties we were known to throw. This was a full ride scholarship paid for by a white guy who lived in Midtown and made a lot of paper. Do you think I cared about that? By the end of the first semester, I was out of there. I would leave on my own accord this time, me being expelled came later. What lead to my sudden departure was a crazy night that carried so much shame I had to take a break and get away, just follow me.

It was a cold December night, around Christmas time. My grandmother's church was having a teens night, and it was actually lit. Girls, food, and hip-hop blasting (I know, what church does that?). What more could one ask for? I attended

The giant wooden barrels (foeders) used to make Rodenbach Flanders Red Ales.
©2012 Christopher Barnes

the event with my right-hand man and linked up with a few other dudes who attended the church. We sat outside in one of the guy's whips, threw back some 211's and E&J, and smoked some weed. Faded & hungry, we went into the church to eat and flirt with the girls. We didn't come in on no fuck shit; we were actually chilling, but little did we know one of the other kids who went to the church brought his homies from Queens...they were on a different type of time.

We ate, laughed, cracked jokes, and talked to the girls while we were blazed out of our minds, not paying attention to the stares and whispers coming from across the room. After a while, we all got on the dance floor, not to dance – just to enjoy the vibe. Next thing we know, one of the dudes walked by and without a sign, swung on my homie. I was high as a kite, and it seemed like everything was in slow motion, like I was watching Neo in the Matrix dodge all those bullets. "Doochie! Doochie!" I heard as the shorty next to me was shaking my arm. "They're fighting!"

I snapped out of my slumber; it's an all-out brawl in the church! I shook off my high and jumped in on the action. Needless to say, the night didn't end well, especially after the ministers broke up the fight and started removing us out of the building. One of the kids from Queens pulled out a pistol on one of the ministers and was really ready to let it off; they really wanted smoke.

Getting kicked out of the church was followed by a lot of disappointment and shame from my grandmother. Not knowing how to manage that, I left NY for six months. Ducked off in Virginia in a new school with the same attitude, I found myself in the same situations. A month into my new school and I was kicked out and placed in an alternative school. That didn't last either. Trouble followed me back then, or should I say, I had a way of finding it. Six months later, I'm back in NYC, and Rice High School was gracious enough to let me back in. My sponsor was also gracious enough to fund my scholarship again, but this was even shorter-lived than the first go around. Within two months, I was expelled. By the first semester of my sophomore year, I had been to three different schools, one twice. This trend didn't stop as I attended eight different high schools in 3 years before I ended up just getting my GED.

Between catching and fighting a couple of felony cases,

Old bottles of 3 Fonteinen Gueuze.
©2014 Christopher Barnes

spending time on Rikers Island, being stabbed multiple times, and just not having a care in the world outside of wanting to be a rapper, school wasn't an option. This attitude took a hard stop when I decided to marry my teen love at 19. Married with a kid on the way, I started to think of shit that had never crossed my mind before, like being alive and free so that I could be involved in the life of my family. I didn't want my children to experience what I did and knew I needed to make a drastic change to make sure that they didn't. In my life growing up, you were either one of two people, a church person who was holier than thou or a street cat. There was no in between, so it's no surprise that in my eyes the only way for me to get myself right was through church.

We immersed ourselves in church and quickly became those people you didn't want to be around – you know, the preachy people who oppose everything. It was a complete 180 from the life I was previously living. This phase of ultra-strict religious beliefs eventually faded, but over the next seven years, we still lived and breathed Christianity. As I began to grow mentally as a man and my mind started to mature, I began to realize that there were more than just two ways to live a good life, it's not only being a street dude or a church person. You could be a loving father, husband, and "model" citizen without being stuck in either extreme or living based off of anyone else's standards; you could set your own. Those thoughts plagued my mind constantly over the last few years of our church stretch.

I know, I know, what in the hell does this have to do with beer? We're getting to that, just follow me. Wifey was pregnant with our fourth and last child at this point. We had already discussed being done, her getting a hysterectomy and us enjoying life with our completed family. On June 7, 2011, our son Josh was born, a healthy and fat baby that brought joy to everyone around him. We were done, we were happy, our family was complete, and wifey went on to get a hysterectomy.

Curveball

August 8, 2011, I'm sitting at my desk at work, and my phone starts going off. Ignoring it due to company policy, I can't help but notice that it continues to go bananas. Finally, I pulled my phone out of my pocket and noticed 911 was calling me.

"Why is 911 calling me?" I thought to myself, and as I answered it, my world turned upside down.

"Mr. Cook," the operator said, "we need you to get home as soon as possible. There's an emergency going on with your baby, and your wife hasn't been able to reach you." I jumped up from my desk so fast and headed to the crib with every negative thought possible flowing through my mind except what I was going to experience when I got home. My son was gone. He was gone. What was a normal daily routine of taking a morning nap didn't end up being normal that day. SIDS is a bitch.

We maintained as best as we could after that, but we were really in mental and emotional shambles. We buried our little homie and decided to celebrate his very short life and the joy that he gave everyone at his funeral. We wanted his siblings to remember him on a good note, but even still...we were emotionally broken. The next few months were hard, and everything came to a head — thoughts of being a good person without extremes, thoughts of life's purpose, thoughts of suicide, thoughts of life being too short and unpredictable not to enjoy it all – the thoughts that were a constant the past few years wrecked my brain. I just needed something to help me cope.

A New Beginning

Scrolling through my kindle looking for inspirational books in hopes of finding something that would help me gain some strength and remain focused, I stumbled upon a book that at first glance of the title made me do a double take. The book was called, "The Search for God and Guinness." Intrigued, I stopped in my tracks, "God and Guinness?" I thought to myself. I mean, I "knew God," but I was a teetotaler, and I knew about Guinness from seeing all the yardies drinking it, but what is this, "The Search for God and Guinness" about? God and Guinness didn't go together – or so I thought. I was genuinely curious.

Upon reading a small sample of the book, something stuck out to me. "I knew I had found it," Author Stephan Mansfield wrote, "that earthy, human, holy tale of a people honing a craft over time and of a family seeking to do good in the world as an offering to God." Just that small sample and I was sold on the book. Just like that, I had to see what this book was about. I purchased the book and dove in immediately, it was so refreshing that I couldn't put it down.

I was in awe the entire time as I learned about the history of this company, everything this family and company was doing was nothing short of amazing. From building homes for employees and the poor, to teaching their employees wives how to read and cook, having medical professionals on staff to take care of employees and their families. From giving employees breaks to read during work, taking care of funeral expenses, to starting the first Sunday school in Ireland – and not in the way that we know of Sunday school today but so that young, poor kids could get an education both in religion and practical matters such as reading, writing, and math. This wasn't even half of the good they did. This multi-million-dollar family company literally changed the world around them through selling beer, all while enjoying it at the same damn time!

Guinness Antwerpen Stout served at Guinness's Open Gate Brewery in Dublin, Ireland. ©2017 Christopher Barnes

I was inspired. I was amped. I was so pumped after I finished this book that I told everyone about it. Guinness had a new loyal fan, and I had yet to ever taste the beer! But oh, was it coming. Due to seven years of seeing the consumption of alcohol as a sin, it took me some time to get my mind right in order to be able to try it; that wasn't an easy mind frame to break.

One afternoon I walked to the bodega and grabbed a single 12-ounce bottle of Guinness Extra Stout, cracked it open and took my first sip… I can't front; it was weird, both me drinking again and the taste of the beer, but for some reason, I wanted to keep taking another sip and another sip until I finished the bottle, and once I did, I went right to sleep and slept good. Ha-ha

After that first experience, I wasn't too pressed to start drinking again. I specifically didn't want to fall back into my old ways, but as I continued to read the book over and over; I always ended up wanting another Guinness. Slowly but surely, I started buying a Guinness a day after work, drinking a beer as I read – it was soothing and healing. At first, the taste had to be acquired, especially for a guy that was sober for so long. I was no longer interested in drinking just to get faded. The taste had to matter for me to drink it, and soon enough, it did.

Shortly after all of this, I left the life of a righteous warrior and started walking in the balance that I knew existed. Although still dealing with the after-effects and pain, my life was a lot lighter and brighter, all because of Arthur Guinness and his heirs. I had hope again. It came from the wildest place, but I dead ass had hope again. I didn't know what it looked like at that moment, but I felt a purpose, too – and some way, somehow, it revolved around beer.

3

BEER IS BEAUTIFUL

Hop cones on the bine. ©2014 Christopher Barnes

From the time I started drinking a Guinness a day, I became more intrigued with beer. So much so, that I started picking up bottles just to read the labels. I read everything from the ABV (alcohol by volume), to how many ounces were in the bottle, to where the beer was brewed and by who, to what style of beer it was all the way down to the ingredients, to when the beer was packaged as well as when it expired. I started noticing that there were different styles of beer and each new one that I tried opened me up even more to what beer really was.

So, let me open you up to the same thing by discussing some of these different styles of beer. Altogether, it's over 100 different styles of beer from all over the world, we can't cover them all here, especially not in depth. So, our list will be concise and simple – but definitely enough for you to get started. I will be referencing style guidelines based off the Beer Judge Certification Program's specifications which will give you general aroma and tasting notes, color of the beer and ABVs. It's important to keep in mind that these are general guidelines.

Beer

Beer is made up of four main ingredients: water, yeast, malt, and hops. Malt is grain that is taken through a process called malting in order to make it useable in brewing. During this process, the grains are transformed into the flavorful backbone of beer. Grain is a crop, grown around the world

Pale and Dark Malt (left). Hop Pellets (right).

and is used for brewing and baking amongst other things. We know grain as cereal, oatmeal, and bread but not as beer. The most frequently used grain in beer brewing is barley, but everything from wheat, rye, oats, as well as many other types you've probably never heard of can be used as well. Each grain has different qualities and effects on the beer's body, the brewing process, aroma and flavor profiles. Aromas and flavors of malt can range from bread to biscuit, from caramel to nutty, to chocolate, to roasted coffee and more.

Hops are another one of beer's main ingredients. If malt is the meat, hops are like the seasoning added to it; they're like Lawry's or Adobo. Let's call them flowers, close relatives to weed but no, hops won't get you high. Many different varieties are grown all over the world; these flowers add bitterness and play as a preservative in beer. They also add aroma and flavor profiles from many different kinds of fruity notes all the way to floral, herbal, earthy and more.

Yeast

Another ingredient, yeast, can be just as important in contributing flavor in beer, but the main job of yeast is to turn sugar water into alcohol. But under the right circumstances, yeast will give off different aroma and flavor notes that can range from fruity ones to peppery spice to some more acquired and or unwelcome aromas and flavors.

Water is no different; depending on location, the quality of water changes drastically and in order to make good beer, the water has to be right. It can add to or take away from the mouthfeel, aroma, and taste if not treated carefully and properly.

Other ingredients can be used in brewing beer: like corn and rice, spices, herbs and fruits, chocolate, syrups and flavorings, all to accomplish different goals that the brewer may have in mind. A lot of different types of grain, hops, yeast and other ingredients all contribute to many different types of beer and beer flavors. Working with all of these ingredients to brew a beer is work but is also a dope experience, let's take a quick look at the process.

An oversimplified look at brewing beer is this: first, the grains are cracked open using a piece of equipment called a mill – this is called milling. They are then soaked in hot water that's set at a target temperature for the specific type of beer they're brewing – this is called mashing. The aim here is to pull out the sugars that are packed inside the grain, these are the sugars that the yeast will turn into alcohol. After the sugar is pulled out of the grains, the water is now called wort, which is just sugar water.

The wort is then collected and boiled for a specific period of time to sterilize and concentrate it. Hops are usually added a couple of times throughout the boil at specific timeframes to add the bitterness, aroma, and flavor "seasoning" profiles. Once the boiling is done, the wort is cooled down, and yeast is added – this is called pitching. The yeast works to turn the sugar water into alcohol, and when their job is done, the beer is ready. Believe me when I say this is oversimplified, tons of science, math, attention to detail, extra steps and skill goes into brewing beer, but this isn't the book for that.

Culture

Lager or Ale

Beer is divided into two categories, Lagers and Ales, with some styles having a foot in both. The differences between these categories come down to a few different things. What type of yeast is being used, what temperature will the beer ferment at and for how long, and the goal of the brewer.

Lagers are intended to be clean beers in taste; yeast can add dozens of different flavors and aromas to a beer, and while that's welcomed in certain styles, in lagers, it's generally not. They are usually fermented at cooler temperatures and for longer periods of time.

Ales, on the other hand, are fair game for the yeast to play. Yeast can add spicy notes like cloves, fruity notes like banana or apple, funky notes like the smells in and around barnyards, and in the right Ale, all of these notes are welcomed with open arms. Ales are fermented at a warmer temperature and are ready to drink much quicker than lagers, usually.

Bottles of Cantillon Lambics (unknown style) conditioning horizantally. ©2014 Christopher Barnes

About Styles

Before we jump into the styles, I'd love to lay some things out for you. There are a few different areas that we pay attention to when it comes to beer. The appearance of a beer is a strong indicator of malt notes that may be present in a beer but should never be used as a judge for the strength of a beer. Concerning malt, lighter beers can display aroma and flavor notes of honey, crackers and bread crust while darker beers can exhibit notes of caramel, nuts, dark chocolate or roasted coffee. Hops and yeast give off an array of different flavors from fruity to spicy to earthy, so unless otherwise specifically stated, these core ingredients are where the flavors in the descriptors emerge from.

Appearance

The different hues that beer takes on can resemble a

rainbow — colors from pale straw to golden, amber, red, reddish brown, to brown all the way to black. All of these colors tell you something different about what you are about to taste. Just as the beer has different colors so does the head (foam) that sits on top of the liquid. These can range from white, thick frothy heads to khaki colored dense heads to light brown, loose heads. And yes, I know it's common not to want any foam in your beer, but it actually is vital to your experience, both aesthetically and as far as the aroma goes. A lot of the aroma that is waiting to be inhaled and enjoyed by you is flowing out of the head like steam off of a cup of freshly brewed coffee. Plus, a good head on the beer keeps the carbonation from going into your stomach and causing bloat and belches, at least a bit.

Aroma

Check this out, our perception of flavor begins with aroma. You ever notice that when your nose is stopped up or when you're sick and can't smell; you also can't taste what you eat? Not to the same point anyway. That's because our sense of taste is closely connected with our sense of smell. We smell in two ways: as we breathe in and as we exhale out, and this sense of smell plays a huge role in our taste. I drive past a bakery every day coming home from work, and the bread smells so good, I could give a damn what it looks like; I want to eat it all – fat boy status. Real talk, the same thing happens when you pass any restaurant, especially if you're hungry. You smell the food and don't even know what it is; it just smells so good that you know you'd bust it down.

But let's flip that script, say your sense of smell was dead, and you only had your eyes to use. The restaurant is serving up the best food in the city, and you're starving, but the food looks like shit; I'm sure it'll be easy for you to pass on it. But with no eyes and only smell, it wouldn't be the same outcome. Our sense of smell is a blessing in and of itself! This is also why we smell our food or drink before we take a bite or sip, even though we've already smelled it! I've heard someone say it before, I don't remember who, but we taste with our nose first.

This is why, with beer, it's important always to smell before you take a sip. It plays a huge role in the enjoyment of your brew. Some of the aromas that you smell may just be in the aroma and not in flavor, and others may be in both, but you'd never know without taking that sniff. It all starts with the sense of smell. Have you ever smelled something and thought to yourself, "that smells too sweet?" I know I have. Also, do you notice that chefs and even you yourself when you are cooking always smell the food on the spoon before you taste it? It's all connected. And as the beer warms up and you continue to take a sniff every now and again, you will notice that even more aromas and flavors start sticking their heads out.

A glass of Westmalle's yeast fresh from the tank. ©2014
Christopher Barnes

Taste

When it comes to taste, each of us can have upwards of 10,000 taste buds in our mouth; some "supertasters" have even more! These taste buds sense five different sensations concerning taste: sweet, sour, salty, bitter, and umami.

Sweet, sour, salty and bitter are all different taste sensations we are very familiar with, we recognize these automatically, and depending on the person, we may either hate some or love them all. In America, it's widely known that we have a sweet tooth, I mean who doesn't love sweet, junk foods? In the words of 50 Cent, "I love you like a fat kid loves cake." And who doesn't love cake? Let's be honest.

We love chocolate, or so we think. What we most consider as chocolate is actually a sweetened version of the original thing called Milk Chocolate. Real chocolate has qualities that are much more bitter, and as I've found out with my children, are not the kind of chocolate they like. Bitter is a taste that we aren't too fond of naturally – especially in America. It's usually scoffed at and considered to be unpleasant to most of our palates, and I get that. Naturally, it's not the most enjoyable sensation in foods or beverages by itself; but when it's balanced out, it's awesome. Coffee, dark chocolate, grapefruit, and many other foods are bitter but nonetheless amazing.

Sour (tart) foods can range from different fruits, yogurts or sour cream – hell, we even have a boatload of different candies that leave our faces deformed while eating them; but if you love them, you love them. Hands in the air for pickle lovers – yeah, I see you, especially ya'll that drink the juice!

When it comes to salty, we love our dishes seasoned, but too much will kill your taste buds and ruin your food, on the other hand, unseasoned food is just plain old nasty, and we all know none of us are going for that shit.

Westmalle Trappist Abbey's copper brew kettle. ©2014 Christopher Barnes

But what about Umami? I'm sure that some of you glossed over that word and didn't think twice. I'm also sure that some of you said to yourself, "what the fuck is that?" Umami is a Japanese word that means "savory," "deliciousness," "yummy." I'm certain that you have experienced this before and just had no clue about what it was. Think eating white rice by itself, now think about white rice smothered in rich gravy and mushrooms...yummy! Think about your spaghetti with just meat and sauce, now think about it with tomatoes and celery chopped up with parmesan cheese sprinkled all over...delicious. One more, think about eating your corner Chinese food with and without the soy sauce, it's a difference – a huge difference.

These are all sensations that our taste buds send to our brain after we taste foods and drinks.

They work hand in hand with our sense of smell to bring us amazing experiences concerning food and drinks and being that beer, in all of its complexity, touches each of these sensations; it's important that we know the ins and outs of them and ourselves.

Mouthfeel

Carbonation

We love drinking soda, not just because of the taste but also for the carbonation that gives us that burp that we need, especially when stuffing our faces and needing a little extra room in our belly. Carbonation is a very important part of beer as well. It's a natural occurrence during the fermentation process. As the yeast converts the sugar to alcohol, it gives off carbon dioxide. That's what gives us that "burp" when we drink soda, and it adds to the overall experience of a beer in every aspect – Including taste. Think of your favorite soda while it has carbonation in it, and now think about how it tastes when it's flat. It's a huge difference, right?

While different styles of beer have different carbonation levels, all beers have it, and it will affect each style differently from taste to mouthfeel. In this sense, beer isn't like soda where carbonation is one size fits all. Beer is way more complex. That tingly feel on your tongue? Carbonation. That burp? Carbonation. Those little bubbles flowing from the bottom of your glass to the top? Carbonation. It's a defining characteristic in beer, from adding to the flavor and mouthfeel to cleaning off your palate while eating, but we will get to that later.

Fullness (Body) & Backend

When we speak of mouthfeel, we are talking of the way the beer feels and flows while in your mouth. The weight of it – does it feel light like water in your mouth or thick and heavy like say a hot chocolate? This coupled with the way it finishes rounds out the mouthfeel. Does it finish dry? Sweet? Tart? Bitter? Does it leave a lingering, very dry feel in your mouth? Or is the finish snappy, nice and quick? All of these things are important to pay attention to and keep note of concerning different styles and your personal preference.

Smell your beer before you taste it and throughout your time drinking it; always pay attention to the flavors and the mouthfeel of it as well as what you like, and dislike about it and why.

4

ALES

STYLES

Now we get to the styles and being that Guinness is where I got my start and is world-renowned for the style of Stout, this is where we shall start. And since stouts are Ales, we will focus on them first and get to Lagers after.

Ales

Stouts

Many people hear the word "stout" and automatically think of Guinness, and that's exactly what should happen. It's a world-renowned brewery that has been brewing beer since 1759, they mastered the style of stout and made it famous. They didn't start out brewing this style of beer, but once they got in their bag, it was a wrap.

Many people also hear the word "stout" and automatically think of a black, heavy, thick, strong beer, and while this could be true, it's not always the case. Stout is a family, so to say. It's a group of beers or "substyles" that fall under the same umbrella. As we stroll through them, you will see what I mean.

Guinness produces three different styles of stout that are easily accessible in the U.S. The Guinness Draught – which is served on nitro, giving it a creamy, smooth body – the Guinness Extra Stout, and the Guinness Foreign Extra Stout. These are three different types of beer with a lot of similarities but some differences as well.

Irish Stout

The first of these is the Irish Stout. The beer usually clocks in between 4-4.5% ABV, with a medium body. A black beer at first glance but when it's really examined, you will notice beautiful ruby rays shining through, especially if held to the light. Roasted flavors are the wave in this beer, notes of coffee abound with other notes of dark chocolate as well. Smooth and creamy, finishing dry with a roasted bitterness on the back end, this beer doesn't fit the stout stereotype.

Suggested Beers: Guinness Draught, O'Hara's Irish Stout, Murphy's Irish Stout

Irish Extra Stout

Irish Extra Stout is like the big brother, not the big, big brother but the big brother who is only a year or two older but thinks he can still boss you around. Yeah, that one. This version clocks in between 5.5-6.5% ABV and is a little more aggressive all around, medium-full to full-bodied mouthfeel; this is a step up. Flavors present here are higher levels of roasted coffee and dark chocolate mingling with notes of caramel and vanilla with a roasted bitterness. A perfect substitute for your cold brew, if I say so myself.

Suggested Beers: Guinness Extra Stout

Foreign/Export Stout

Again, we find ourselves going a bit bigger. The Foreign Extra Stout is bigger than the previous two in every way. Clocking in between 6.3-8% ABV, it's a beer that isn't brewed to drink like water. This beer continues on the theme of the roasted coffee and chocolate notes, but it mixes in a little more sweetness in the aroma with some fruitiness present as well as some other notes like molasses. The beer is medium-full to full bodied with a moderate carbonation level, dry finish, and higher bitterness than the previous stouts.

Suggested Beers: Guinness Foreign Extra Stout, Pike XXXXX Extra Stout

Tropical Stout

Tropical stouts are similar to Foreign Extra stouts, but differences are present. This is my favorite style of stout, and one I frequently dream about drinking while sitting at a local hole in the wall restaurant in the Caribbean stuffing my face.

Quick story: one day as I was working the bar slinging beers in hot ass Florida, a group walked in. One of the people in the group asked for my recommendation, and I suggested a stout, they went on a stereotypical rant of how you can't drink them in the heat; they're too thick and heavy. After letting them say their peace, I started to explain that Tropical Stouts and Foreign Extra/Export stouts are frequently enjoyed in the Caribbean. They have been exported to tropical islands for decades, and they're also brewed there now because they're staples for a lot of the locals. I don't know if it was the heat or the fact that they didn't care to hear what I had to say, but the level of big mad they were was hilarious.

Tropical stouts start off sweeter than all the others mentioned so far. This beer is still dark with notes of coffee and chocolate, but those notes are much more chill. Higher notes of apples and bananas with a medium to full body are the vibe here. This beer also has a quality that will put you in the mind of rum, and being that it's a Caribbean staple, is that surprising? 5.5-8% ABV and a restrained bitterness. This beer is fire.

Suggested Beers: Lion Stout, Dragon Stout

Milk Stout

Milk stout, which is also known as cream stout, is sweeter, creamier and less roast forward than the other stouts. The roast is still present; bitterness is here, also, but it's balanced by the lactose (the sugar in milk) used in the beer. If some of the other stouts are like enjoying black coffee or coffee with very little add-ons, a milk stout is your standard coffee with cream. Fuller feeling mouthfeel, creamy, sweet, notes of vanilla and chocolate and coffee all melded together. This beer usually sits around 4-6% ABV and is definitely a pleasure to drink.

Suggested Beers: Left Hand Milk Stout, Left Hand Milk Stout Nitro, Mackeson's XXX Stout

Oatmeal Stout

Oatmeal stouts can have a similar sweetness and feel as milk stouts, or they could be drier. It all depends on the brewery. The oats add a creamy mouthfeel and some flair to the roast that this family of beers is known for. Fuller bodied, medium bitterness – at 4.2-5.9% ABV, this is a damn fine choice to choose from.

Suggested Beers: Samuel Smith Oatmeal Stout

Samuel Smith Oatmeal Stout

Russian Imperial Stout

These are the big boy stouts. Sitting anywhere between 8-12% ABV, these beers are sippers. Thick, chewy, full-bodied beers with roasted and burnt notes, licorice, molasses, dark fruits like raisins, prunes and plums as well as caramel, chocolate, and coffee. These are beers to be sipped slowly. They can range from English variants that are mild in the way of bitterness to American variants that are highly bitter.

Samuel Smith Imperial Stout

Suggested Beers: Samuel Smith Imperial Stout (English), Great Divide Yeti, North Coast Old Rasputin, Sierra Nevada Narwhal (seasonal), Lagunitas Imperial Stout, Bells Expedition Stout, Oskar Blues Ten Fidy

Other Stouts

There are many other types of stouts available that you will cross as you continue on this journey. From oyster stouts (yes, oyster shells and even oysters are used), to stouts flavored with chili peppers, cacao, Nutella, French vanilla, if you name it, it has probably been done. Mexican hot chocolate stouts, pastry stouts – which are usually very thick and sweet, flavored with anything the brewer desires. Fruited stouts, barrel aged stouts that are aged in every kind of barrel from whiskey to bourbon to tequila. The options available are overwhelming.

Suggested Beers: Sierra Nevada Stout, Brooklyn Brewery Black Ops, Ballast Point Victory At Sea, Bells Black Note, Prairie Artisan Ales Bomb, Stone Xocoveza (seasonal), Goose Island Bourbon County (seasonal)

Porters

Bell's Porter

If you're not careful as a new drinker, you can easily mistake Porters for Stouts due to their similarities but be aware of their differences. Porters are like Stout's baby brother, light brown to dark brown in color with ABV ranging from 4-6.5%, depending on whether it's an English or American version. English versions will showcase a restrained roasted and bitter character displaying aroma and flavor notes of nuttiness, caramel, and chocolate with floral or earthy hops, medium bodied with a creamy character and moderate carbonation. American versions, on the other hand, will be more aggressive. Dark malts still may showcase the chocolate and even coffee with a medium body and moderate carbonation, but the biggest difference is the role that the hops play. The flavor profiles of the hops can be resiny, floral or earthy in character and may be more in your face as opposed to the subtler hops in English Porters.

Suggested Beers:

English – Fuller's London Porter, Samuel Smith Taddy Porter, Green Man Porter

American – Anchor Porter, Sierra Nevada Porter, Deschutes Black Butte Porter

These beers are at their best served between 45-55 degrees as the colder these beers are the more the flavor and aroma profiles will be disguised. My advice, let it warm up so that the beer can open up and shine like it's supposed to.

Wheat Beers

Opening myself up to new beers and flavor experiences was a great decision that I'm happy I made once I became aware of what was available to me. The next three beers that I experienced were what are known as Wheat beers. This is another family of styles with several substyles, some of those styles even have their own family within them (see Lambic). These beers are made using barley but also have a large percentage of wheat used in the brewing process. The most common theme in wheat beers is that they all use wheat, outside of that, the final product is usually very different, unlike stouts.

Wittekerke Wit, brewed by Brouwerij De Brabandere. ©2015 Christopher Barnes

Witbier/Biere Blanche: White Beer

Originating in Belgium, this beer is one of the first styles of beer that I tried after Guinness. Usually pale to light gold and has a honey sweetness from the malt. Subtle notes of coriander playing with oranges and other citrus notes with a medium body. Highly carbonated with a refreshing zest and a lower hop bitterness with earthy notes. Clocking in around 4.5-5.5%, this beer is often brewed with orange peels, coriander, grains of paradise and other spices. Depending on the brewery, some may be sweeter, some may have a tart finish, but it's definitely a great beer that's full of life.

Suggested Beers: Allagash White, Cigar City Florida Cracker, Blanche de Bruxelles, Timmermans Lambicus Blanche, Brewery Ommegang Witte, St. Bernardus Witbier, Wittekerke, New Belgium Fat Tire BelgianWhite Ale

Hefeweizen

A well respected but overlooked beer style that's been around for ages. Hefeweizen (also known as Weisse or Weizen) is a German beer with a rich history. This is the wheat beer of wheat beers — golden color with a haziness that keeps you from seeing straight through. This beer sits between 4.3-5.6% and is a highly carbonated, highly refreshing beer with low bitterness. Fluffy, creamy mouthfeel, this is a beer to be enjoyed outside on a beautiful day. Although we taste the bready notes from the wheat, the yeast is the star of the show here. Bananas and cloves wrapped in a beautiful display, even bubble gum and vanilla have been known to make appearances.

Weihenstephan Hefeweisse Dunkel

Darker and stronger versions (Dunkelweizen, Weizenbock) exist and will add different depths to the flavors from the yeast-based off what malt is used, but banana, cloves, and toasted bread are always welcome in my cup any day.

Suggested Beers: Weihenstephaner Hefeweissbier, Paulaner Hefeweizen, Ayinger Bräu Weisse, G. Schneider Original Weisse, G. Schneider Aventinus (Weizenbock), Sierra Nevada Kellerweis, Weihenstephaner Vitus Weizenbock

Tart and Sour Wheat Beers

This group of wheat beers has either a tartness or a mouth puckering quality to them. Come on, don't act like you don't like lemonade or don't enjoy sour patch kids. These beers are true gems and pleasures to enjoy, and once you are over your shock, you'll thank yourself and me for trying them. In the midst of some stand-alone styles we also have a family with multiple substyles within this group, just ride with me and pucker up.

Gose

Gose, pronounced (Goes-zuh) is a beer that sits around 4.2-4.8%. It's brewed with sea salt and coriander. The salt adds to the mouthfeel, and the spice makes it pop. Light and bright, the sourness is kept in check and fruits like lemons, apples and pears come out to play, balanced by bready notes from the wheat with low bitterness. A dry, medium bodied beer that's highly carbonated and every sip leaves you wanting another, it's refreshing AF, too. Versions with fruit added are very popular with American breweries.

Ritterguts Gose

Suggested Beers: Anderson Valley Gose, Döllnitzer Ritterguts Gose, Westboro Gose, Bahnhof Leipziger Gose

Berliner Weisse

A Berliner Weiss has more of a sour kick than a Gose but sits at a much lower ABV, usually no more than 4%. This beer is also lighter bodied but displays similar notes as far as light fruits and grainy bread. Its carbonation is high, the mouthfeel is light yet juicy, and this pale beer is overall refreshing and easy drinking.

Suggested Beers: Prof. Fritz Briem 1809 Berliner Weisse, Bahnhof Berliner Weisse, Mikkeller Hallo Ich Bin Berliner Weisse (this beer comes fruited in a handful of varieties)

• • •

Today, these beers are, a lot of times, "fruited," meaning that brewers are adding in fruits to give them a twist or to kick up the experience, but this wasn't always the case. Syrups like raspberry and woodruff, used to be mixed in leaving these yellow beers either red or green with a balance of sweet and tart. Woodruff is my favorite to add, hands down. It leaves your beer tasting like my favorite childhood cereal, Lucky Charms, others have said Apple Jacks.

Lambics: Lambic, Gueuze, Fruit Lambic

A vintage bottle of 3 Fonteinen Oude Gueuze served at the 3 Fonteinen Cafe. ©2014 Christopher Barnes

The family of Lambic beers are a highly praised and respected group of beers hailing from Belgium; these beers are special and highly regarded. Brewed then aged in oak for upwards of three years and sometimes longer, then blended with newer and older beers until the artist believes that they have the perfect product. Yeast isn't added to these beers, nope, the wort is left to fend for itself while the elements of nature provide the yeast that will transform it into beer. This is why these beers are known as "spontaneously fermented." It's all left up to the environment, the atmosphere, just like it was when beer was first stumbled upon and for thousands of years after until we discovered what yeast was and how it worked.

Hops are not used in these beers as a seasoning, they're only wanted for one thing, and that's to help protect the beers. Here, they want the preservative aspect of hops to be the number one focus. These beers can be known to have some wild flavors and aromas from the wild yeast, and, why shouldn't they? They're wild beers.

As we talk Lambics, you will see the word "oude" mentioned in some of the beer names. Oude means "old" and it refers to the beers being made according to tradition as they have been for decades. These beers will show-case tartier and funkier notes than ones without oude in the title.

Oud Beersel Oude Lambiek served at a Brussels cafe. ©2014 Christopher Barnes

Lambic

This is the base beer all the following beers are created from. This yellow to golden colored beer is usually hard to find in the states but every now and again you just may get your hands on one – I have before, and it was official. It's the only one in this family of beers that is not blended and is served as is – straight. Uncarbonated, the flavors of apples, lemons, and other citrus flavors are side by side with a mouth puckering tartness as well as notes of earthiness, hay, goat, and barnyard.

I told you they were wild! I can see you looking at the book like, "what the fuck?" But trust me, if you are familiar with cheese and the aromas associated with them but know how good they are, then this is exactly the same. A lighter bodied beer that swaps bitterness with sourness. Sitting at 5-6.5% ABV, these beers can also be aged for years and enjoyed like vintage wine. You can sometimes find draft versions at bars that know what's up.

Suggested Beers: Oud Beersel Oude Lambiek, Boon Oude Lambik

Fruit Lambic

Krieks (cherry), Framboise (raspberry), Pomme (Apple) and a lot of other fruited lambics are easy to get your hands on. These beers are aged in oak with whatever type of fruit the brewer/blender decides on. This light to medium bodied beer is carbonated and blended with 2-3 different vintages to produce a masterpiece. Usually, no higher than 7% alcohol, these beers still showcase their wild side alongside an extra appearance of fruit. Some are sweet, and the puckering sourness is no longer

3 Fonteinen Oude Kriek served at the brewery.
©2015 Christopher Barnes

noticeable while some are still tart as ever with just a fruitier pop.

Suggested Beers: St. Louis Framboise, Lindemans Framboise, Lindemans Kriek, Lindemans Cassis, Boon Framboise, Boon Oude Kriek Mariage Parfait, Boon Kriek, Oud Beersel Oude Kriek, St. Louis Fond Tradition Kriek, Lindemans Cuvee René Oude Kriek, Tilquin Quetsche (plum)

Gueuze

Gueuze, pronounced "Goo-zuh," is a golden colored, light bodied, highly carbonated blend of different vintages of lambic. Mouth puckering, wild notes with light, bright fruits all intertwined together with honey and bready malts. Topping off at 8% ABV, this crisp beer finishes dry, and although its sour quality is the center of attention, it's a highly balanced beer.

Suggested Beers: Lindemans Cuvée René, Oud Beersel Oude Geuze, Boon Oude Geuze, Boon Oud Geuze Mariage Parfait, De Troch Oude Gueuze, Tilquin Oude Gueuze

These beers are usually bottled in 375ml and 750ml champagne-style bottles, caged and corked, and are able to be aged and enjoyed at the moments you desire. Be careful, treat these bottles like Champagne! Or you might just pop yourself in the face with a cork.

Wheat beers are best enjoyed at temperatures between 40-50 degrees with family, friends, and foods. Just like stouts, there are other styles of wheat beers available that I didn't touch on here but as you explore, I'm sure you will come across more variants and styles from this family of beers.

Bottles of Tilquin Quetsche (plum), Tilquin Oude Gueuze, &
Tilquin Gueuze(Squared). ©2015 Christopher Barnes

Flanders Reds and Oud Bruins

This group of beers is like first cousins that are totally different. Both originated in Belgium, Flanders Red in West Flanders while Oud Bruins came out of East Flanders, think side by side like Brooklyn and Queens. These also fall into the scope of sour beers, but they aren't wheat beers. Although, they are like cousins in a sense; their upbringing is totally different – Brooklyn and Queens.

Flanders Red

Flanders Red Ales are Red to brown in color with fruity notes like cherries, flavors of chocolate and vanilla and a sourness balanced out by soft malt. These beers don't focus on hop bitterness just like the other sours, but some light bitterness may still be noticed. They have been called the "red wine of beer," an elegant drink that comes in between 4.6-6.5%. These beers are also aged in oak for prolonged periods of time and blended like lambics. A medium bodied, dry, fruity, sour ale.

Rodenbach Grand Cru served at a cafe in Roeselare, Belgium. ©2012 Christopher Barnes

Suggested Beers: Rodenbach Alexander, Rodenbach Grand Cru, Rodenbach Vintage, Rodenbach Fruitage, Duchesse de Bourgogne, Vichtenaar Flemish Ale, Petrus Aged Pale (brewed in Flanders like a Red, but pale in color)

Oud Bruins

Oud Bruins are darker in color compared to Flanders Red Ales and tend to be more malt focused than sour. Although a low sourness can be tasted in this beer, it is usually not as sour as any of the others. Medium bodied with notes of caramel and toffee, plums and raisins, chocolate and cherries. The bitterness in this beer is restrained like a Flanders Red. This is also a blended beer; however, it isn't aged in oak like its cousin. ABV ranges from 4-8%. Oud Bruins, as well as Flanders Reds, can be and sometimes are aged on different fruits.

Suggested Beers: Liefmans Oud Bruin, Liefmans Goudenband, Petrus Oud Bruin

Just like lambics, these are beers that you can age and enjoy for years to come. Serve at the same temperature as wheat beers.

Fantôme Saison and Chocolate Saison served at the brewery in Belgium. ©2014 Christopher Barnes

Saison

Also referred to as Farmhouse Ales due to the history of the style. These are usually gold in color, but darker ones are also available. They are light to medium bodied, highly carbonated beers with ABV's ranging from around 3-9%. They are full of flavor, too! Spicy, herbal and floral notes from the hops, peppery spice from the yeast mixed with notes of citrus and balanced bitterness. This ale was known to be made with whatever was lying around so grains of all kinds could be employed in the beer and make experiencing this style truly unique. Serve around 45 degrees and drink up.

Suggested Beers: Saison Dupont, Brooklyn Sorachi Ace, Boulevard Tank 7, Fantôme Saison, Dupont Avec Les Bons Voeux, Saison d'Erpe-Mere

Trappist & Abbey Ales

Trappist Ales are beers made by religious monks. Yes, they make beer. In all of my time being involved in religion, I had no clue! Oh, what we hide to push our agendas. These deeply religious monks have been brewing beer for a very long time, it's how they take care of their personal needs as well as the needs of the monastery. I already know the next question so let me go ahead and answer it. Yes, they drink beer as well. Although the beer that they drink is said to be lower in alcohol than some of the ones they sell, they still enjoy beer. In order to be considered Trappist one must have the approval of the "secret council," and their beer must carry the sacred seal, "Authentic Trappist Product."

A Trappist Monk of the Rochefort Abbey in Belgium pouring some of his Abbey's beers. ©2014 Christopher Barnes

Just joking about the secret council, but you do have to be a member of the International Trappist Association and meet strict criteria to use it. There are 12 Trappist breweries in the world today, and their beer is some of the best brewed. On the flip side, Abbey ales are beers that are brewed at non-religious breweries with licensing from the churches they are working with. These church orders are usually different than the Trappist, but as is with their religion, their beer is similar at the same time. Yeast is a big player in these beers; peppery, spicy notes are prevalent. Oh, and as well as a lot of notes of fruit and candi sugar.

Single

This medium bodied, highly carbonated, yellow to darker colored beer sits between 4.8-6%. Sweet malts, fruit aromas of different types from oranges to apples to peaches mixed with spicy & floral hop notes and spicy, peppery notes from the yeast all blend well in this bitter beer that finishes dry.

Suggested Beers:
Trappist – Rochefort 6, Chimay Dorée.
Secular — St. Bernardus Extra 4 (Seasonal), St. Bernardus Pater 6.

Dubbel

The dubbel is a rich beer, dark amber to copper to brown in color. This beer is a little higher in alcohol, usually between 6-7.6%. It packs a medium to full body with good carbonation but is pleasantly smooth. The bitterness here is lower as is the spice since the focus is on the malt.

St Bernardus Extra 4.
©2015 Christopher Barnes

Caramel, chocolate, and toast are all possible notes in these beers as well as dark fruits and cherries — a beer that's chalked full of flavor with a dry finish.

Suggested Beers:
Trappist — Chimay Première (Red), La Trappe Dubbel, Rochefort 8, Westmalle Dubbel
Secular — St. Bernardus Prior 8

Westmalle Dubbel. @2014
Christopher Barnes

Tripel

Tripels take the ABV and bitterness a bit higher. These yellow to golden beers are highly carbonated and medium bodied. These beers are deceitful, easy to drink and without paying attention, they will sit you on your ass. 7.5-9.5% the alcohol in these is well hidden, and you'd definitely think it's much lower without checking — malty sweet like honey with similar tasting notes as the single just more pronounced.

Suggested Beers:
Trappist — Chimay Cinq Cents (White), La Trappe Tripel, Westmalle Tripel
    Abbey — Val-Dieu Triple
    Secular — St. Bernardus Tripel, Straffe Hendrik Brugs Tripel, Kasteel Tripel

Quad

This is the big boy of these beers. Known as Quads, Quadruples, or Belgian Dark Strong Ales, these beers are sippers like Russian Imperial Stouts. 8-12% ABV, they are dark amber to brown. The beer is like a dubbel on steroids, bigger body, and richer.

Brouwerij Van Honsebrouck's Kasteel Triple. ©2015 Christopher Barnes

Suggested Beers:
Trappist — Chimay Grande Réserve (Blue), Achel Extra Brune
    Secular — St. Bernardus Abt 12, Straffe Hendrik Quadrupel

Other Suggested Belgian Beers: Orval, Piraat, Duvel, Delirium Tremens, Gulden Draak

The lighter of these beers do great served between 40-45 degrees while the darker ones are best between 50-55 degrees.

Chimay Grand Reserve sampled at the Abbey in Chimay, Belgium. ©2017 Christopher Barnes

Some Other Ales

The next set of beers can come in many different varieties and interpretations – from English to American versions to more malt focused ones to more bitter ones. Normally, if it's from England, it's not going to be as bitter as ones brewed in America. That's not always the case, but it generally is, with these beers and with every other style as well.

Amber Ales

Amber ales are usually between 4.5-6.2%; these beers are usually amber to brown in color with a medium body and high carbonation. Flavors can include caramel malty sweetness, citrus, floral, piney, tropical, melon or whatever other hop flavor the brewer is shooting for. Usually, a balanced beer between bitterness and malt but that too depends on the brewer.

Suggested Beers: Cigar City Tocobaga Red Ale, Anderson Valley Boont Amber Ale, New Belgium Fat Tire Amber Ale

Brown Ales

Cigar City Maduro Brown Ale

Brown ales are a favorite of mine. Chocolate, caramel, nutty, toffee, biscuit flavors are all possible. Mixed with low floral, herbal or woody flavors if it's English or high fruity flavors if it's American. 4.2-6.2% in ABV, this medium bodied beer can have a low to pronounced bitterness.

Suggested Beers:

English — Samuel Smith Nut Brown Ale

American — Brooklyn Brown Ale, Bells Best Brown Ale, Cigar City Maduro Brown Ale, Florida Ave Brown Ale, Avery Ellie's Brown Ale, Big Sky Moose Drool Brown Ale

Pale Ale and IPA

Pale Ales and IPA's have been popular for a long time, and they're great styles so I can see why. These beers come in many different takes and ranges of alcohol. Usually golden to amber in color, these beers are more focused on having the aroma and bitter qualities of the hops at the forefront. Flavors and aromas have a broad spectrum from citrus or piney to tropical fruits to melons or berries; the list is pretty broad.

Pale Ales are lower in ABV and everything else overall. IPA's pack more of a punch in each aspect of the beer, from ABV to bitterness, flavor, aroma, you name it. IPA's are also available in specialty versions: Red IPA, Belgian IPA, Brown IPA, NEIPA or Hazy IPA, etc. Most of these variants are like hybrids of two styles, take a brown IPA for example. The malt flavor of a brown ale with the high bitterness levels and hop forward profiles of an IPA is what you'll get. Swap the brown ale malt with the spice from the yeast of a Belgian beer, and a Belgian IPA it is.

MIA Mega Mix Pale Ale

NEIPAs, also known as Hazy IPAs, are different. They break the rules of IPA's. They focus on all of the flavor and aroma from the hops minus the bitterness. They are very soft and non-aggressive. Each has their place and serves a different purpose, each is good, it just depends on what vibe you're feeling.

Suggested Beers:

Pale Ale — Sierra Nevada Pale Ale, Cigar City Guyabera, Anchor Steam, Brewers' Pale Ale, Oskar Blues Dale's Pale Ale, Deschutes Mirror Pond Pale Ale

IPA — Bells Two Hearted, Lagunitas IPA, Fat Heads Head Hunter IPA, Sunshine City IPA, Cigar City Jai Alai, Hardywood Park The Great Return IPA, New Belgium Ranger IPA, Sierra Nevada Torpedo Extra IPA, Deschutes Fresh Squeezed IPA

Brown IPA — Dogfish Head Indian Brown

I would suggest serving the Amber, Brown, Pale Ale and IPA's between 40-50.

Highly Suggested Beers: Two styles that we did not touch on but are highly suggested for you to seek out. Fuller's ESB, Fuller's London Pride

Barleywines

A big, beastly beer that's always around my crib. These big beers are sippers, and I usually bring them out on chill days. Think of lounging, laid up all day watching movies – Netflix and chilling with a Barleywine. I also like to enjoy them in small amounts, 6-8-ounce pours when possible after a big dinner. These beers range from highly bitter (American versions) to malty, sweet, and rich (English versions). They are also beers that can do well aging, given the right environment. These beers can range in color from blonde to copper-brown with alcohol levels from 8-12 plus percent. Some may be aged in wine or liquor barrels while some are not. Thick and chewy beers. Malty sweet with caramel and toffee, fruity notes and sometimes candy, hop aromas, and flavors are whatever the brewer decides. Lower carbonation than a lot of other beers, this is a beer to take your time enjoying. Serve between 50-55 degrees.

Suggested Beers:

American — Anchor Old Foghorn, Sierra Nevada Bigfoot, Great Divide Old Ruffian, Lagunitas Gnarleywine, Brooklyn Brewery Hand & Seal

English — Fuller's Vintage Ale, J.W. Lee's Vintage Harvest Ale

5

5

LAGERS

STYLES

Lagers

Lagers are clean beers. A lot of the fruity and spice forward flavors mentioned with the ales are not present in lagers. Lagers are malt and hops all day, beers that are flavorful but at the same time don't have a lot going on like Ales, they're straightforward. A common mistake people make is looking at lagers in a general way. Even new drinkers of craft beer automatically think of Corona, Heineken, Miller Lite, or Budweiser to be the face of all lagers. Yes, these beers are lagers, but these beers do not represent what a lager is even in the slightest. That's like saying one of these new age mumble rappers represents the entirety of hip-hop. Cut it out; it just doesn't make any sense. Continue to follow me, and you'll get the picture.

Pilsner

Pilsner is the OG that most of the above-mentioned beers are subpar replicas of. Birthed out of the Czech Republic, it was the first beer of its kind, and when it was released in the mid-1800s, it set the world on fire. Either people wanted to mimic it as best they could, or they were challenged to make better beers that they felt represented their cities and regions instead of allowing a foreign beer to take over. It's two different kinds of Pilsners that we will touch on here, Czech and German Pilsners.

Czech and German Pilsner

Uneticky 10° Czech Lager served at their restaurant outside Prague. ©2015 Christopher Barnes

The Czechs are known to have multiple styles of beer that are well regarded; they're also known to drink the most beer in the world. It's an everyday part of their lives, enjoyed with meals, family, and friends. The beer that you're most likely to get your hands on here is their premium lager. The one that started it all, Pilsner Urquell. Many breweries in the states take their shots at brewing both types so I'd suggest that you pay close attention to what category the beer falls under if you come across one.

The Czech Pilsner is golden in color with a medium body and low carbonation. Rich, bready malts with subtle notes of spicy, floral or herbal hops with a noticeable but soft bitterness and rounded finish, at 4.2-5.8% these beers are very refreshing and easy to kill.

The German Pilsner is a little different. The body is lighter, the color is lighter, and the carbonation is higher. Low notes of grainy malt reminiscent of crackers mixed with higher floral, herbal or spicy hop aromas and flavors. The ABV ranges from 4.4-5.2%- this beer finishes crisper and drier than its Czech counterpart. 38-45 degrees is where these beers are at their best.

Suggested Beers

Czech — Pilsner Urquell, Praga Premium Pils, Lagunitas Pils, Budvar, Rebel

German — Paulaner Premium Pils, Victory Prima Pils, Bitburger Pils, Flensburger Pilsener, Sierra Nevada Nooner Pilsner, Jever Pilsener

Helles

This yellow to gold lager comes in between 4.7-5.4% ABV with a clean aroma of grainy malt sweetness and spicy, floral or herbal notes from the hops. The flavor profile follows suit with a grainy, malty sweet impression and noticeable but non-aggressive hop flavors. Medium bodied with medium carbonation, this beer finishes soft and dry. A simple yet flavorful beer that is easily a daily go to.

Suggested Beers: Paulaner Premium Lager, Hacker-Pschorr Münchner Gold, Spaten Premium Lager, Weihenstephaner Original, Cigar City Tampa Style Lager, Augustiner Bräu Edelstof, Hofbrau Original, Andechs Vollbier Hell

Schwarzbier

This beer is a black lager – the name Schwarzbier means, black beer. People quite often mistake the color of a beer with how heavy or strong it is, but as I hope you have realized by now, that's just not the truth. Schwarzbier is a treat. 4.4-5.4% ABV with a medium-light body, this beer is the right beer anytime. Moderate carbonation, brown in color with a nice smoothness that's astounding. This beer is roasty with notes of unsweetened chocolate and spicy, herbal or floral hop flavors.

Suggested Beers: Köstritzer Schwarzbier, Mönchshof Schwarzbier

By now, I'm sure that you're noticing that lagers are simple and straightforward in taste, that they are, but they're also the hardest beers for a brewer to make with no issues.

Dunkels

Dunkels are copper to brown in color with a medium body and soft, rich mouthfeel with medium carbonation. 4.5-5.6% with malt flavors of toasted bread, nuts, caramel and maybe even chocolate. Hop bitterness is typically lower in these, but the herbal, spicy, floral notes can still be present.

Suggested Beers: Ayinger Altabairisch Dunkel, Hacker-Pschorr Alt Munich Dunkel, Weltenburger Barock Dunkel

Both of these dark lagers are best served at 45-50 degrees.

Bocks

Bocks are a family of beers under one umbrella like Lambics. These beers come in lighter and darker colors with flavor profiles that match the grains used – from toasted notes to rich bready notes to caramel. Herbal, floral or spicy hop notes stand out more in the lighter colored beers than the darker ones. ABV, depending on which style of bock you choose, can range from 6.3% to 14 plus percent. A few of the different styles of bock are: Dunkel Bock, Maibock, Doppelbock, Eisbock (which is pronounced Ice-bock) 45-55 degrees is perfect for these beers, the higher the ABV, the more you should let it warm up.

Suggested Beers:

Maibock — Ayinger Maibock, Hofbräu Maibock, Einbecker Mai-Ur-Bock.

Doppelbock — Ayinger Celebrator, Paulaner Salvator, Spaten Optimator, Altenmünster Winterbier Doppelbock, Augustiner Maximator, Andechs Doppelbock.

Eisbock — G. Schneider Aventinus Eisbock, Kulmbacher Eisbock

Hofbrau Maibock

Märzen

Märzen is a creamy, medium bodied beer with medium carbonation that finishes dry. 5.8-6.3% ABV, this beer is amber to copper and tastes of bready toast, malt sweetness with lower floral, spicy, herbal hop notes and moderate bitterness.

Suggested Beers: Paulaner Oktoberfest, Hacker-Pschorr Original Oktoberfest, Weltenburger Kloster Anno 1050

Paulaner Oktoberfest Marzen

Hybrids

Hybrid beers are beers that have one foot in each category. It can be because the beer uses an ale yeast and ferments at lager temperatures or vice versa. Needless to say, they are the best of both worlds.

Altbier

Altbier comes in between 4.3-5.5% and is a beautiful amber to copper color with a medium body and medium-high carbonation. Its bitterness is a little aggressive, but that's fine. It's dry and crisp with cherry aromas from the yeast that may be perceived as well as some nuttiness alongside low notes of spicy, peppery or floral notes from the hops.

Suggested Beers: Uerige Classic, Uerige Sticke, Uerige Doppelsticke

Check a local brewery or beer store to see if any of these or others are available

Kölsch

Reissdorf Kölsch

Kölsch, which is hands down in my top three styles of beer, is light and refreshing and crisp. 4.4-5.2% ABV, medium-light body with medium-high carbonation, easy to drink with a dry finish. Originating in Cologne, Germany, this golden ale has medium hop notes of floral or herbal character with light honey bread mingling with aromas of apples, cherries or pears. This simple, yet delicious ale is a year-round classic for me and one that's sure to leave you refreshed.

Suggested Beers:

German — Früh Kölsch, Gaffel Kölsch, Sünner Kölsch, Reissdorf Kölsch.

American — Left Hand Travelin' Light Kolsch, Boulevard American Kolsch

Best served at 38-45 degrees if you want to get the most from these beers.

What's important to note is that these brief descriptions based off the Beer Judge guidelines are what you can usually expect from these styles of beer in general. These are what are known as traditional descriptions of these beers, but each brewer has a different thought in mind when they are designing their recipes, so subtle or big differences can occur. With a few questions to the bartender, store clerk or even reading the label you should be able to have a clear understanding of what you're getting. Never be afraid to ask a question, when it comes to beer, we are all still learning no matter how someone may act, so ask away, in confidence.

6

EAT, DRINK, AND ENJOY

What Beer is Right for Me?

This isn't hard to answer, especially if you ask yourself the right questions. Like, what kind of foods do you like? Do you like sweets? Do you have a knack for things that are sour? What about sweet and sour combinations? Are you a fan of bitter things? Are salty foods your vibe? What kind of drinks do you enjoy? Both alcoholic and nonalcoholic.

Do you love the bitter and roasted taste of coffee with no cream and sugar? Or do you have to drown your coffee in sugar? Do you like tea? How do you like yours? Do you enjoy dry wines or sweet wines? Do you even like wine? Do you like your hard liquor straight? Mixed? Sweet? Balanced? With bitters? Do you like dark chocolate or milk chocolate?

These are all questions you should be asking yourself, especially since you now have a quick break down of a lot of different styles of beer to choose from. I mean, unless the place you purchased it from just sucks at making beer, or unless it's a bigger issue like a spoiled beer, you'd win at picking beers way more than not with asking yourself these questions.

I love watching shows with hosts like Andrew Zimmern and the late Anthony Bourdain, Rest in Peace, big dog. They encourage you to try different things; Andrew even encouraged us to try everything twice. I'm saying, how do you know that you won't like it if you don't try it? You don't! And even if you didn't

Bavik Super Pils and Beef Carpaccio.
©2015 Christopher Barnes

like something in the past, that doesn't mean you won't like it now. The biggest thing that holds one up from fresh and dope experiences is a closed mind. Maybe it was that one bad experience with some nasty ass beer that we shouldn't have been drinking in the first place that robbed us of experiencing great moments like a jack boy on the prowl. Maybe it's just that you never tasted beer but your outlook on it is that it's a bottom shelf drink (you thought). It could be any of these things or a little of them all mixed together that holds you back.

Sometimes when we try new things we automatically love them; some things have to grow on us as we challenge and train our palates intentionally. Other times, our palates change on their own as we get older. When I was a kid, I hated string beans; my mother had to force me to eat them. Now, as an adult, I don't know what made me hate them that bad as a kid. Maybe being forced to eat them back then made them grow on me or maybe my palate just changed with time, the latter has happened plenty of times with me, for sure. That's a fact.

Before I got back into drinking beer, I loved buying those cappuccinos out of 7-11 — the ones that you get out of the machine that's basically just powdered sugar mixed with hot water. Damn, I used to love those! I couldn't stand coffee any other way. Now, since I've been drinking so many different beers, I don't want any sugar in my coffee. Just the thought of buying one of those cappuccinos makes me nauseous. I'm all the way good on that. This has happened time and time again with multiple foods and beverages, and it's all because our palate changes whether we are intentional or unintentional about it. Either way, I encourage you to challenge your palate and always seek to expand it, let it try things that it never has and watch out because, in no time, you'll be eating and drinking shit you would have never considered.

Beer and Food

While we are talking about trying new things, we may as well tackle beer and food. Now, I'm no Garrett Oliver – he's a very smart guy when it comes to beer and food, and you can grab the brother's book, "The Brew Master's Table," to fully immerse yourself in this topic if you'd like – but what I do know is that my food experiences drastically changed once I began to properly pair the right beer with the right foods. Beer and food paired properly are like Jadakiss and Styles P going back and forth on a Swizz track in the early 2000s, it's unmatched. Nothing beats the right dish with the right brew, and this goes for all foods. Breakfast, lunch, brunch, dinner, snacks, appetizers, desserts, cheeses, you name it; there's a beer that'll be perfect for it.

A fairly simple benchmark that I learned in my studies was the concept of the three C's. The three C's are Complementing, Contrasting, and Cutting. You can complement the flavors of a dish with the flavors of a beer. Like herb crusted chicken with herb spiced potatoes and veggies paired with a Belgian Blond Ale. The spices of each complementing one another from beginning to end. Or we could complement flavors by drinking a stout while eating a slice of rich chocolate cake; it's really better with the stout than it is with milk.

We can contrast by putting flavors against one another, like sweet and sour. We can take that same piece of sweet chocolate cake and drink a tart fruit beer with it.

Cutting is when we are eating rich foods and need something to slice through the richness. Cutting can come from the bitterness, carbona- tion, or the alcohol in the beer, each will cut through fatty, greasy and tongue coating foods with no issue.

A great meal pairing experi- ence that I recently enjoyed was shrimp and grits with a Saison. We all know that grits are thick and coarse in texture, these were made with cheese and different herbs and spices and

Boon Oude Geuze and Pheasant Roulade at the world famous Hommelhof Restaurant in Watou, Belgium. ©2014 Christopher Barnes

mixed with the shrimp sprinkled with different herbs and spices. It was amazing, but definitely a rich meal. The Saison was the perfect drink for it at that moment, light citrus flavors that tasted like I had squeezed a lemon over top of my shrimp mixed with the spices of the beer that complemented the ones in the dish. The high carbonation and alcohol in the beer cut through the thick, rich grainy character of the grits and cheese with no prob- lem. It was A1!

The thing to be mindful of in all of these instances is not overpowering either the beer or the food. You don't want the food or the beer to be overshadowed, so keep this in mind, a big beer more than likely won't fare well with a light meal. No

Russian Imperial Stouts with a Caesar salad! That's a stretch, but I've seen people do it! Throw some steak, strong tasting veggies, and some thick dressing on it, and we can start to look at bigger beers, but still not imperial stouts.

It's also important to look at all of the different flavors in a beer as well as the dish that you're thinking about because just as a beer can complement the food, it can also clash with it. One not so great experience that I had was pairing a tropical stout with some rich banana, blueberry and peach oatmeal. My young reasoning was, "Well, the stout is rich and has fruit flavors that will go well with the flavors in the oatmeal, and the oatmeal is thick and rich, so they should be able to handle each other." What I didn't think about was the "rum qualities" that were present in the tropical stout, and while the fruits in the oatmeal did mesh well with the beer, the rum quality clashed with everything else. It was an epic fail, but a lesson learned. We all take L's sometimes, stop cappin' like you haven't.

Jamaican Red Snapper served with a witbier. ©Dom "Doochie" Cook

Beer is also constantly used as an ingredient in food, and I know Steel did it with the 40oz in the movie "Juice," but we're better than that. No 40s of malt liquor in our eggs, man. I love cheesecake, if you don't then now, I have to worry about you. But I really love my cheesecake drenched in cherries and sauce, it's something I've loved since I was a kid. So, a Kriek, especially one of the sweeter variants, would be a no-brainer to pair with it, right? But what if I made my own cherry sauce with that same Kriek to drizzle over my dessert? Winning! We made a cheesecake, grabbed some cherries and a bottle of Lindemans

Kriek and got to work! The outcome was a dope ass dessert. Cheesecake smothered in a kriek lambic sauce. Now, that's the only way to have it. Another dope dessert we love to make is a chocolate cake from scratch with stout as one of the main ingredients, this adds a richer depth and flavor profile of chocolate to the cake and feels like a fireworks show is happening on your palate. It's a beautiful experience.

As you get to know more flavor profiles of different beers, you can experiment more with pairing beer with food and with using beer as an ingredient. If you are a chef, whether professionally or you just do the damn thing at your crib, and you really know food, then this will be great for you to explore. Just keep in mind, it's a marathon. I'm sure that there are times when you think of a great idea recipe wise, and it doesn't come out the way you envisioned, no? But when that does happen what do you do? You tweak it until its right. It's the same with beer. Don't be discouraged, all of these different flavors in beer are new to you, and it can be a lot to grasp but I know, and I think you know, that you got this. And if you want extra help, there are different beer cookbooks available to give you a starting point on this journey.

St Benardus Prior 8 and game bird at Hommelhof Restaurant. ©2014 Christopher Barnes

The last angle to look at concerning beer and food pairing is this, each style of beer has a country, city, and region that it hails from with their own customs and traditions concerning beer and food, and a lot of times, they have their own set of rules they follow. It's been working for them forever so who are we to fuck tradition up – especially if the pairings they eat day in and day out are solid? West Indian food with a stout, for example, is a no-brainer. We even have our own pairings here in America like pizza and pilsners, they're great together, but that's not the only beer that pairs well with a slice. You should always be down to try new things with different beers and food but also experience the joy that other cultures and their eating and drinking traditions bring as well. Now excuse me as I finish this beef patty with coco bread and stew chicken with my export stout while I fantasize about the tropics.

7

RESPECT BEER

APPRECIATION & MODERATION

Appreciation is defined as the recognition and enjoyment of the good qualities of someone or something. I'm sure there are a lot of people or things in your life that you appreciate. We recognize the good that these people or things bring into our lives, and we enjoy them and the moments we spend with them. I'm also sure beer has never crossed your mind as something to appreciate, tell me I'm wrong.

In ancient civilization, they appreciated beer. From the time that the first person came across this strange drink, it made such a huge impact that it turned their world upside down. If archeologists are correct, it was appreciated so much so that it made them finally decide to settle down in one place. It made them change their entire way of living. Their appreciation of beer was also shown in their religious beliefs. Gods and Goddesses were dedicated to the "nectar of the gods" known as beer.

Sumer, which was one of the first two civilizations that we know of, sat where southern Iraq sits currently. It's believed that this is where most of our technological advances were birthed from; it's also believed that this is where beer got its start. This isn't saying that this is where humanity started, no, it's saying that this was one of the first times that humans who lived nomadic lifestyles, moving from place to place, decided to settle down and build cities. The Sumerians were a religious group of people; they worshipped multiple gods/goddesses that all had different purposes to fulfill. So, it should come as no surprise that they worshipped one dedicated to beer. Ninkasi was the goddess who satisfied the heart with beer and bread. She was a brewer to the gods and the people, a guardian of beer, one who brought lots of joy.

Ninkasi being a goddess and in charge of beer is also an early example of the role women have played in the beverage. Since the beginning, women have brewed beer and brewed it well for the enjoyment of the home and also for the financial aspects of taking care of the household. Women are pivotal in beer: past, present, and future and should be respected as such.

Ancient Egypt is the other early civilization that had a strong appreciation for beer. Inscriptions about beer in the tombs of mummies inform us of one thing, the people wanted to enjoy it even in the afterlife. I can't blame them for that; I want to do the same. Beer was so important and appreciated that workers even accepted payment in liquid form. The appreciation of this beverage had to be to the utmost. The Egyptians, who also had many gods, worshipped Osiris in regard to beer and brewing. He was an important god known for a lot of things, and the gift of barley and beer was amongst them.

Tribes throughout Africa have brewed and drank beer forever at celebrations, at religious events and just in everyday life. It's a beverage that many civilizations throughout time have had gods and goddesses set aside for, and if this seems interesting to you and you'd love to learn more, I'd encourage you to do so. It's a beverage that has been viewed as a daily part of life for many different groups of people for ages. One that was indulged in, appreciated and enjoyed daily with high regard. This brings me to my next point: Moderation.

Orval Trappist Ale freshly bottled at the Abbey brewery in Belgium. ©2015 Christopher Barnes

Moderation is simply the avoidance of excess or extremes. This is something that we should aim to practice in many areas of our lives but especially with beer. Since I started drinking with appreciation, my view has been, "a beer a day keeps the doctor away." Usually, two beers, depending on the ABV is a great medium for me to enjoy what's in my glass and still have my wits about me. Now don't get me wrong, there have been more times than I care to recall of going overboard, but thankfully that isn't the norm.

Moderation is important; life is about balance and part of balance is being able to control your impulses and desires. Beer is great, it's a gift that we have been given to enjoy, but it has also been a factor in causing many people and families great harm because someone didn't believe in setting limits. It's not about the buzz, although that's a nice part of drinking. It's definitely not about getting drunk; hangovers fucking suck. It's about appreciating this flavorful drink in front of you and the experiences you have while enjoying it. Simple pleasures, that's the most overlooked gem in the world. Solomon, the king of Jerusalem, spoke repeatedly about eating, drinking and enjoying life; these are the focuses I believe we should have when it comes to life but especially with beer. Eat good food with people you love, drink good beer with people you love, and enjoy it all, for this is truly a gift from God.

8

IT'S BIGGER THAN BEER

LONG RUN HUSTLE

"I had hope again, I didn't know what it looked like at that moment, but I felt a purpose, too, and some way, somehow it revolved around beer. "

You remember this from earlier? Everything comes full circle. I had no clue at that moment where beer would take me or what it would allow me to accomplish, but boy was I in for a ride. The first few years I tried every beer I came across, I read every article I could find, I obsessed over beer daily. Not in my drinking habits, at least not for the most part but definitely in my thoughts. I left New York for Virginia, and my beer journey continued.

It was there that I realized that I hadn't even scratched the surface with beer; it was there that I came across my first store dedicated strictly to beer. It was also there that I came across bars dedicated to the same. It was there that I realized that local breweries were a thing! It's not that NYC didn't have any of these things, because they did. It's just that as I stated before, the focus of these establishments was not on the urban demo-graphic, and I was unaware they existed. You either stumbled upon it by chance or someone who has, like me, exposes you to it. I was hyped to see stores and bars and breweries dedicated to beer but, one thing constantly bothered me – every beer store that I visited, every beer bar or brewery that I had a drink in, I was always the only black guy there. Due to social media, I came across a few blacks scattered across the country who loved good beer but in real life, none. This was always unacceptable in my eyes and always something that needed to be changed.

For the next three years, I continued to taste, read, observe and soak up as much as I could. I loved beer and never wanted to play with this craft; I always wanted to study under and learn from the best. I always wanted to constantly expand  my knowledge, I mean, everyone I came across had an opinion on beer, but the issue was that most of them had no authority to stand on. It was too many critics without enough credentials. Through observation, I learned early on that to be respected as a voice concerning beer one had to have the proper credentials and or experience to back them up, you had to put the work in, something that I took to heart and ran with.

In the wine world, authority and respect are given to a person who is certified as a Sommelier. These are wine stewards who through experience, education, and exams become voices of authority concerning wine. In the beer world, they are called a Certified Cicerone®; these are individuals who through experience, education, and exams become the same – respected individuals concerning beer. The Cicerone website describes these individuals as someone who "possesses the knowledge and skills to guide those interested in beer culture, including its historic and artistic aspects." I never graduated from high school (I got my GED, remember) or college, but I knew that this was the "higher education" that I was destined for.

Beer was only a hobby at the time, one that I wanted to eventually turn into a career, someway, somehow and after much thought and consideration, I decided to take the leap. I left a consistent, secure and well-paying job, packed my family's bags and headed to a young but thriving beer scene in Central Florida. The craft beer revolution, as it's known in the United States, got its start in the '70s but started to grow in the '80s and really caught wind in the '90s.

Breweries popped up all over the country, but some cities and regions were a little slower to catch on than others, Tampa was one of them. Although the oldest craft brewery in the state of Florida is based in Tampa Bay and a few others followed suit, it really didn't take off until around 2009-2010. This scene was still a baby yet growing rapidly and showing signs of greatness. That mixed with great weather and palm trees was a no-brainer for us. It was here that I decided to carve my name in the cement, it would just prove to be harder to do than anticipated.

It took two years of hard work and dedication, but I accomplished my goal of earning my Certified Cicerone® certification as well as another title from the Beer Judge Certification Program. These were and still are huge feats in my eyes. With only around 3,600 Cicerones in the world, I was one of them, and I was proud of that. But even with knowledge of higher pursuits in specialized training, my road to making beer a career was rough. Once in Florida, I faced rejection after rejection, 40 breweries took a look at me and said, "no" for entry-level positions. It was a rough time indeed, but my love for beer and the support of wifey helped me continue to push, and through the blessings of a dear brother, I caught a break.

I got hired and started at a brewery as a part-time tour guide – at the bottom. Working one to two days a week while surviving off a "real job" which I hated. Every day I dreamed of only working in the beer industry, and in due time, that dream became my reality. I loved what I did, even if it was being a tour guide. I was usually the first to show up and the last to leave. I always went above and beyond and was always ready to get the job done. So, it came as no surprise when the

Kozel Lager served at a Prague Cafe.
©2015 Christopher Barnes

then manager of the bar, now my close friend, offered me a shot as a bartender. He offered me the shot as I came in to get ready for a tour one morning, and he needed me to start directly after the tour. I was happy but scared; I had never bartended a day in my life! I, however, took the opportunity and ran with it. Fulfilling two different positions in the brewery allowed me to leave the job I hated; what a time to be alive!

After about five months of bartending, I was promoted to assistant general manager as well as being tapped to help out with a lot of other things around the brewery, and I must admit, it was a blessing to be living this life that felt like a dream. Since then, I've held other positions within the beer industry, but my mind has always been stuck on my kulture. I've met a lot of great people in this industry, both professionals and consumers, from different walks of life. I've made some great friendships along the way as well, but what I haven't been able to come to grips with is that the demographic that looks like me is not being exposed to good beer – we are still being bombarded with the bullshit. It was rare working in the beer industry to see another black; there was literally only a handful in the Tampa Bay Area when I started. It was even rarer working the bar and seeing blacks come in to enjoy a beer, definitely not cool.

G. Schneider Aventinus Eisbock

An underlying mind frame that I've heard too many times to recount is that blacks won't like flavorful beer. Another is that it's a luxury that we won't or can't spend money on. This, coming from industry shot-callers who make the decisions but have never put these theories to the test. Only garbage beer has been marketed to us, so how would anyone know what we will or won't like? These mind frames infuriate me and are part of what spurred me to write this book – someone has to step up to the plate.

"Good beer for all" is the slogan. That's also the reality that I desire to see lived out in and around this industry – minorities from all backgrounds working in the industry as well as being consumers of great beer. This is an industry that's still growing and providing trades, jobs, education, and business opportunities, and it's one we should have our hands and feet in while remaining true to ourselves and being who we are.

Beer Can Change the World Again

Even bigger than the previously stated, I've mentioned how beer has been used for the betterment of humanity by Guinness. Arthur and his family really put work in to make their city and the world around them a better place – all through beer! They were in the trenches doing some real shit with the resources that they acquired through brewing. Shit that you wouldn't expect to see done by a brewery or multi-million-dollar company. Period.

I mean, imagine Colt 45 actually being on the ground taking care of the poor in the craziest parts of the city. Imagine St. Ides investing in children who aren't privileged or receiving the best education because of their zip code or the color of their skin. Imagine that. Imagine a company giving enough of a fuck that they hire a doctor to take care of their employees and their families, minus the expensive ass co-pay charges of insurance. But wait…. Picture a beer company that actually fought hard to right a lot of the wrongs we see in this world. Picture it. Who knew that through beer you could do so much good? The Monks did & so did Arthur.

The ability to use your career, business, money or even beer to change the world is still very attainable. Beer may have a bad rap because of some people's overindulgence and abuse of it, but with the right attitude and mind frame, we can see it initiate positive change as it has before — something that our world desperately needs at the moment. That could look like you opening a bar or brewery in your neighborhood and providing jobs to the residents there, teaching the trade of brewing to others so that they have something that they can survive off of under their belts. How many people do you know without a trade or much work experience? I know quite a few myself.

*Guinness West Indies Porter served at the Guinness Open Gate
Brewery in Dublin, Ireland. ©2017 Christopher Barnes*

Using beer to change the world looks like keeping the money in our communities and building up where we are through reinvesting in ourselves and our brothers and sisters – especially the ones coming behind us. It can look like giving back to the less fortunate, starting a "nonprofit" branch to your business, providing a familial type of environment for families and friends to kick it in peace where our children can learn the importance of appreciation and moderation.

There are countless ways that we can invest and give back to our community. Whether it's falling in love with beer and wanting to use it as your vehicle or if it's another means of transportation you take, as long as it's getting done, it's all good. This is the purpose, not just exposing you to beer that you never knew existed but also to show you how to handle it responsibly as well as using my platform and resources to do good and inspire others to do the same.

9

THIS AIN'T THE BEER THAT YOU'RE USED TO

As stated at the beginning of this book, beer is much more than what we've been told it was, and I hope that I have painted a clear enough picture for you to see that. Beer's diversity in taste, aroma, and appearance is vast and beautiful. Its history is even wider and too rich for little old me to get into in this book. There's always a new style to try, and always more knowledge about beer to learn. There's always a new memory to make with friends and family around beer. As you taste and take mental notes of your likes and dislikes, keep three things at the forefront of your mind: one, always keep an open mind and two, be aware that your tastes usually change over time on their own, so why not aim to intentionally expand your palate by challenging it? Three, don't get caught up in the glamour of what's cool or in, drink what you truly enjoy.

Beer is a drink to be enjoyed, not abused but enjoyed, just as life is. It's best enjoyed with family and friends, with food or after a hard day's work when you need to unwind. So, if you have that perfect beer already that fulfills these things and you are content and satisfied with that, keep on enjoying your beer, whatever kind it may be. But if you're not content with that beer that you have in your hand or in the fridge, and you believe more is available then fuck with me. Let this book be the starting point for your new adventure of tasting new beers. Let it be your companion to the store or bar while you're picking up bottles and cans and reading the labels. Let it be the starting point on your journey, but never let it be the ending point. The information and education available to help teach you and show you even more is readily available in this day and age.

Whether you find beers that you like and just want to enjoy them in your free time or whether your new-found passion pushes you to start home brewing or even to make a career out of brewing or working in another part of the beer industry, you can do it; all are available options to you. The world is yours, and the continued growth and success of this beer industry depends on people like you and me. This is just the beginning; this book wasn't meant to be exhaustive or your final stop. My story doesn't stop here, and neither does my learning and growth, and yours shouldn't either. Be great, be fearless, stay dangerous, for the Kulture! Cheers

Love is love,

My Beer Gangsta is Respected, B

ADDITIONAL READING & SOURCES

Recommended Reading

- *The Oxford Companion to Beer*, Garrett Oliver (editor)
- *The Brew Masters Table*, Garrett Oliver
- *Beer Judge Certification Style Guidelines*, BJCP Program
- T*he Beer Bible*, Jeff Alworth

Educational Programs

- Cicerone Certification Program
- Beer Judge Certification Program
- Siebel Institute

Sources

Books

- Jackson, Michael. *The New World Guide To Beer.* Philadelphia: Running Press, 1997
- Mosher, Randy. *Radical Brewing.* Boulder, CO: Brewers Publications, 2004
- Mosher, Randy. *Tasting Beer.* North Adams, MA: Storey Publishing, 2009.
- Oliver, Garrett, ed. *The Oxford Companion To Beer.* New York: Oxford University Press, 2012.
- Oliver, Garrett. *The Brewmaster's Table.* New York: HarperCollins, 2003.

Web Articles

- GeoGee (2016). *How Menthol, Malt Liquor, And Black Folks Became Entwined.*
- Godlaski, Theodore M. (2011). *Osiris of Bread and Beer.* https://www.researchgate.net/publication/51237807_Osiris_of_Bread_and_Beer
- *McKenzie River Corporation (2018).* Retrieved from https://en.wikipedia.org/wiki/McKenzie_River_Corporation
- Mark, Joshua J. (2011). *The Hymn To Ninkasi, Goddess Of Beer.* Retrieved from https://www.ancient.eu/article/222/the-hymn-to-ninkasi-goddess-of-beer/
- Trager, Louis (1996). *Lager of Legend.* Retrieved from

https://www.sfgate.com/business/article/LAGER-OF-LEGEND-3149914.php

- Winship, Kihm (2012). *Malt Liquor: A History.* Retrieved from https://faithfulreaders.com/2012/04/29/malt-liquor-a-history/

ABOUT THE AUTHOR

Dom "Doochie" Cook is the co founder of Beer Kulture, a company that specializes in bringing awareness and exposure of good beer to the urban landscape. He is a Certified Cicerone® and BJCP beer judge with three years experience working in the beer industry. He loves studying about beer and seeing the excitement on the faces of people who never thought that they could like beer after he shows them that this ain't the beer that they're used to. His hope is to help paint a new picture of what beer is, what it isn't, and what it could be for a new generation of drinkers.

facebook.com/beerkulture

twitter.com/beerkulture

instagram.com/beerkulture

Printed in Great Britain
by Amazon